IEEE Radio and Antenna Days of the Indian Ocean (IEEE RADIO 2019)

IOP Conference Series: Materials Science and Engineering
Volume 766

Saint-Gilles, Belgium
23 - 26 September 2019

Editor:

Vikass Monebhurrun

ISBN: 978-1-7138-2664-4
ISSN: 1757-8981

Printed from e-media with permission by:

Curran Associates, Inc.
57 Morehouse Lane
Red Hook, NY 12571

Some format issues inherent in the e-media version may also appear in this print version.

Institute of Physics
Dirac House, Temple Back
Bristol BS1 6BE UK

Phone: 44 1 17 929 7481
Fax: 44 1 17 920 0979

techtracking@iop.org

Additional copies of this publication are available from:

Curran Associates, Inc.
57 Morehouse Lane
Red Hook, NY 12571 USA
Phone: 845-758-0400
Fax: 845-758-2633
Email: curran@proceedings.com
Web: www.proceedings.com

TABLE OF CONTENTS

IOP Conf. Series: Materials Science and Engineering **766** (2020) 011001 doi:10.1088/1757-899X/766/1/011001

Editorial Board Members

IOP Conf. Series: Materials Science and Engineering **766** (2020) 011002 doi:10.1088/1757-899X/766/1/011002

EDITOR'S FOREWORD

The 2019 edition of the IEEE Radio and Antenna Days of the Indian Ocean (IEEE RADIO 2019) international conference was hosted in Reunion Island for the second time. This seventh edition of the conference was held from 23rd to 26th September 2019 at the Lux Resort & Hotel, Saint-Gilles Les Bain. The conference was managed by the Radio Society (Mauritius) and financially sponsored by the IEEE Antennas and Propagation Society (AP-S). Région Réunion financially supported the conference and Union Radio Scientifique Internationale (URSI). provided technical co-sponsorship. The Local Organising Committee consisted of researchers from CentraleSupélec, France, University of Mauritius and University of Reunion Island.

IEEE RADIO 2019 was attended by participants from about twenty different countries. The conference featured a tutorial on the design of radio amateur networks followed by presentations of radio amateur activities in Mauritius and Reunion Island. Projects related to humanitarian activities were discussed during a special session organized by the IEEE Special Interest Group on Humanitarian Technology (SIGHT) Indian Ocean. Two workshops, open to all the participants of the conference, were organized in conjunction with the conference. The first workshop addressed measurements and simulations for industrial applications and development of standards. The second workshop was dedicated to the CARERC (Cartographie Electromagnétique par Réseaux de Capteurs) project that was undertaken by colleagues at University of Reunion Island during the past years. The regular sessions covered topics such antenna design, bioelectromagnetics, devices and circuits, wireless communication and tracking systems. Three internationally recognized scientists delivered invited talks during the conference.

The oral presentations of students and young scientists were evaluated by a jury. A shield and a cash prize of 300 Euros were awarded to the Best Young Scientist less than 36 years old. The presenters of the first, second and third Best Student Papers were awarded shields and cash prizes of 300 Euros, 200 Euros and 100 Euros, respectively. The AP-S sponsored Industrial Engineering Paper Award, consisting of a shield and a cash prize of 300 Euros, was presented to the author of the paper presenting the most innovative and significant contribution in terms of practical industrial engineering related to the fields of antennas, electromagnetics and propagation.

All the 2-page short papers that were effectively presented during the conference have already been published on the IEEE Xplore Digital Library. A call for the submission of 4-page extended versions of the conference papers was launched for publication as a volume in the IoP Conference Series: Materials and Engineering (MSE). After the peer-review process by the distinguished members of the Scientific Committee, only four papers were accepted and published in this volume of IoP Conference Series: MSE.

Editor
Vikass Monebhurrun

IOP Conf. Series: Materials Science and Engineering **766** (2020) 011003 doi:10.1088/1757-899X/766/1/011003

Peer review statement

All papers published in this volume of *IOP Conference Series: Materials Science and Engineering* have been peer reviewed through processes administered by the proceedings Editors. Reviews were conducted by expert referees to the professional and scientific standards expected of a proceedings journal published by IOP Publishing.

IOP Conf. Series: Materials Science and Engineering **766** (2020) 012001 doi:10.1088/1757-899X/766/1/012001

EM Modelling Considerations for mm-Wave On-Chip Antennas

Flavien Sagouo Minko and Tinus Stander

Carl and Emily Fuchs Institute for Microelectronics
Dept. EEC Engineering, University of Pretoria, 0002, Pretoria

tinus.stander@up.ac.za

Abstract—We investigate the inaccuracy of resonant frequency (f_0) estimation incurred by some common simplifications in electromagnetic (EM) simulation of mm-wave on-chip antennas. Using an 88 GHz patch antenna as a device under test (DUT), it is found that using a 2.5D planar EM solver produced highly inaccurate results. Including the bulk silicon and thin nitride in the model stack effected an improvement of 0.8% in estimating f_0, while replacing individually modelled via arrays with equivalent blocks, and accounting for the etch factor, marginally improved both the estimation of f_0 and reflection coefficient $|S_{11}|$. It is shown that, with the correct modelling choices, the error in f_0 can be reduced to below 50 MHz at E-band, which is within the measurement variation of 880 MHz.

Index Terms—**Finite element analysis, millimetre-wave integrated circuits, numerical simulation, patch antennas.**

1. Introduction

Millimeter-wave on-chip antennas present many challenges to full system-on-chip (SoC) integration [1], including low gain and large footprint. Both problems can be assessed accurately through reliable first-iteration full-wave electromagnetic (EM) modelling [2], [3]. Full-wave 3D EM modelling of large on-chip passive devices is complicated by the extreme aspect ratios and thin dielectric layers, both of which contribute to large mesh sizes. The high frequency also exacerbates the losses associated with surface roughness and the etch factor. This complex modelling environment often necessitates simplifications of the model geometry [4], such as 2.5D (as opposed to full 3D) modelling[5], the omission of probe pads [6] and of some dielectric layers [7]. The inevitable effect of these simplifications is the reduced first-iteration accuracy, with a deviation of several hundred MHz [7] for the resonant frequency (f_0) of resonant patch antennas above 60 GHz.

In this paper, we investigate a number of simplifications used in the EM modelling of mm-wave on-chip antennas and quantify how each affects the accuracy estimation of f_0 in resonant patch antennas implemented in the back-end-of-line (BEOL) of a typical BiCMOS process. In addition to the effect of BEOL stack first conducted in [8], this work investigates the effect of via geometry simplification, etch factor and solver choice.

2. Device under test

A simple square patch antenna (figure 1)[8] was fabricated in BEOL of the GlobalFoundries 8HP SiGe BiCMOS process (7 metal layer BEOL). The 818 × 818 μm patch is located on the top thick (±4 μm) (AM) aluminium layer and suspended over a large, solid copper ground plane (1070 × 938 μm) on MQ layer at a height of ±9.6 μm below AM. The patch is fed with an AM-MQ microstrip line of width 12.54 μm and length 200 μm, which is excited by ground-signal-ground (GSG) probe arrangement of 100 μm pads and 150 μm pitch. The ground pads are stitched to the large MQ ground plane with via array stacks of 39 × 39 μm.

IOP Conf. Series: Materials Science and Engineering **766** (2020) 012001 doi:10.1088/1757-899X/766/1/012001

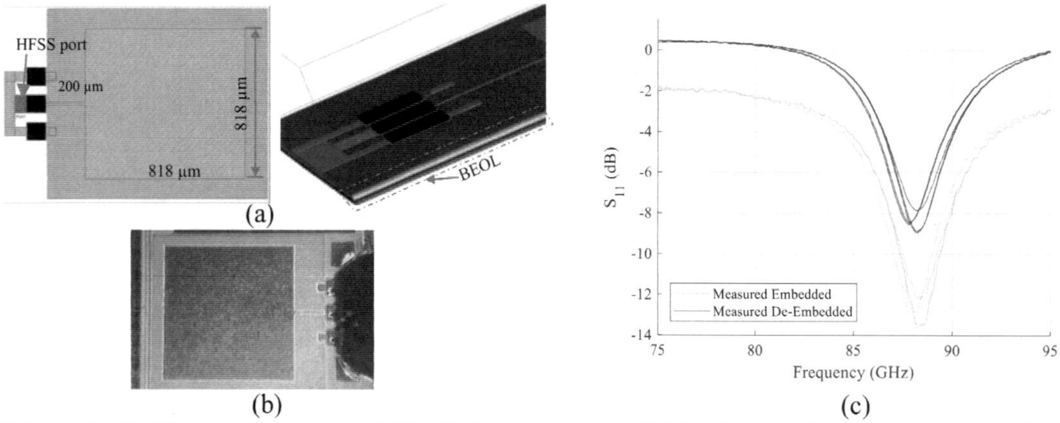

Figure 1. 88 GHz patch antenna: a) HFSS simulation model. b) Micrograph of the antenna under test. c) Embedded and de-embedded measurement results

The device under test (DUT)'s S_{11} is measured through wafer-probing, with three devices measured to establish measurement error margins. Both probe-tip calibration using line-reflect-match (LRM) standards on a ceramic substrate and second-tier de-embedded using on-chip thru-reflect-line (TRL) standards results are shown in figure 1(c), where the measured f_0 ranges between 87.72 to 88.6 GHz and $|S_{11(min)}|$ from -13.69 dB to -12.12 dB in the embedded case. Due to a discrepancy in the definition of the THRU and REFLECT standards on-chip, a minor de-embedding error results in $|S_{11}|$ = +0.33 dB out-of-band. With this result established, different EM modelling approaches were investigated.

3. EM modelling comparisons

3.1. Effects of stack-up and pads

Three versions of the 7ML BEOL stack-up (using the 3D layout design procedure in [2]) were investigated (figure 2). Stack "A" omits the FEOL and terminates on the lowest M1 metal routing layer. Stacks "B" and "C" both include the bulk silicon terminated in a perfect electric conductor (PEC) and heterogeneous SiO_2 across the BEOL. Stack "C" also includes a thin nitride ($\varepsilon_r = \pm 6.8$) layer above the bulk Si substrate as opposed to an oxide layer ($\varepsilon_r \pm 4$), in stack "B". The results in figure 2(d) indicate marginal variation between stacks "B" and "C", both for embedded and de-embedded measurements, while stack "A" may be disregarded in the further analysis due to the significant discrepancy. The background plots in grey represent the measured results.

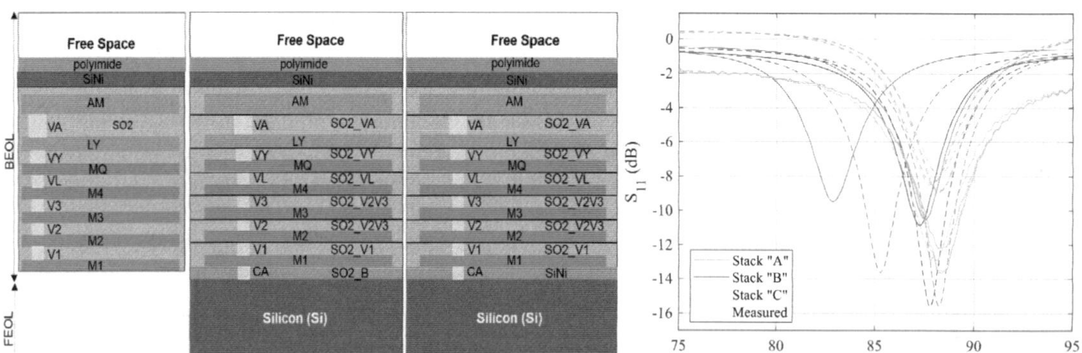

Figure 2. Comparison of BEOL stacks : (a) Stack "A", (b) Stack "B", (c) Stack "C", (d) Simulated results, with embedded pads are shown in solid lines and de-embedded results with dashed lines[8].

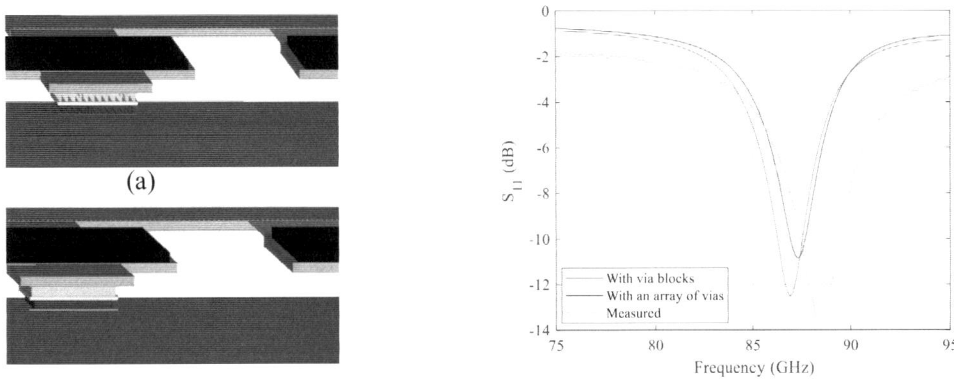

(b) (c)

Figure 3. Via block simplification: (a) Via array modelling, (b) Consolidated vias, (c) Results

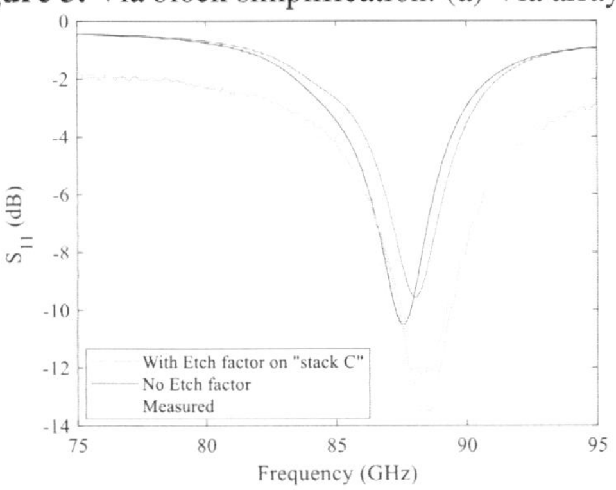

Figure 4. Etch factor comparison

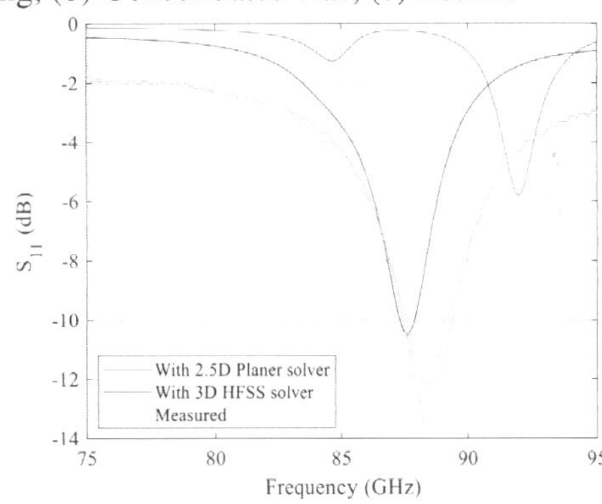

Figure 5. Solver comparison

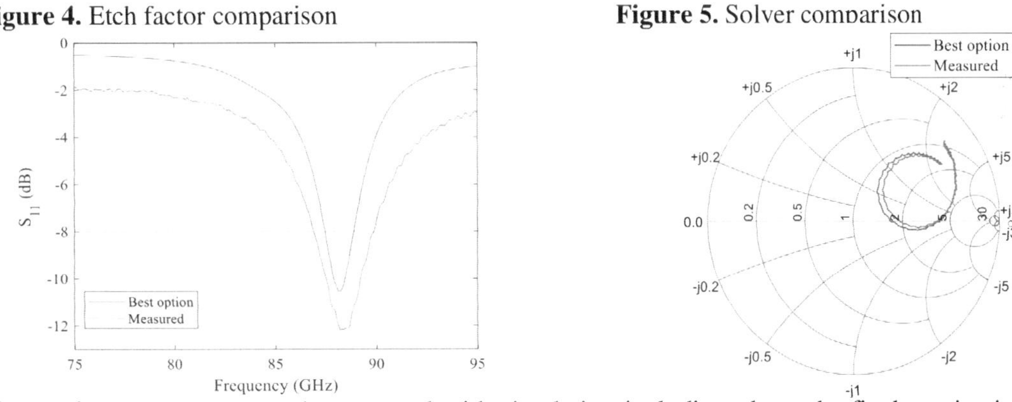

Figure 6: Measurement result compared with simulation including, the pads, final passivation, via blocks and etch factor

3.2. Effect of via simplification

To evaluate via simplification effect on the modelling accuracy, solid via blocks (figure 3(b)) are used instead of arrays of vias, using stack "B" and embedded pads. The results in figure 3(c) indicate no considerable loss of accuracy in estimating either f_0 or $|S_{11(min)}|$, as the resulting changes of 87.3 MHz and -12.49 dB are within the measurement error range as demonstrated in figure 1(c).

3.3. Effect of etch factor

Using stack "C", the effect of etch tapering [9] is investigated by including and omitting a 71° etch angle. The resulting variation of 360 MHz in f_0 and 0.35 dB in $|S_{11(min)}|$ shown in figure 4, indicates very little effect on modelling accuracy when compared to the measurement error in figure 1(c).

3.4. Effect of solver choice

Using stack "C", 2.5D planar and 3D finite elements method (FEM) solvers were also evaluated with identical simulation settings. The comparison in figure 5 clearly indicates the inaccuracy incurred in using this approach, despite the advantages in reduced solver time [3].

3.5. Consolidated selections

By using Stack "C", with embedded pads, with 71° etch taper, consolidated vias and FEM solver, the comparison in figure 6 is obtained. Less than 50 MHz and 1.7 dB are incurred in estimating f_0 and $|S_{11(min)}|$, respectively.

4. Conclusion

We have presented a comparison of different EM modelling approaches using an 88 GHz patch antenna in BiCMOS as DUT. It was found that the thin-metal approximation in the 2.5D planar solver yielded significant inaccuracy, while a FEM solver could estimate f_0 and $|S_{11(min)}|$ accurately to within 50 MHz (0.056% error) and 1.7 dB when the bulk silicon, consolidated via arrays, and the etch factor are included in the EM model. Future work will extend this analysis to an investigation of the antenna's radiation pattern.

Acknowledgements

The authors wish to thank the MOSIS Educational Programme (MEP) for sponsoring prototyping space in the GFUS 8HP process. This work was further supported by the National Research Foundation of South Africa (NRF) under Grants 92526 and 93921, as well as the Eskom Tertiary Education Support Programme (TESP).

References

[1] H. M. Cheema and A. Shamim, 'The last barrier: on-chip antennas', *IEEE Microw. Mag.*, vol. 14, no. 1, pp. 79–91, Jan. 2013, doi: 10.1109/MMM.2012.2226542.

[2] T. Stander, 'A novel, flexible Pcell-driven design flow for mm- wave on-chip passives', in *IEEE International Microwave and RF Conference (IMaRC)*, 2015, pp. 209–212.

[3] F. S. Minko and T. Stander, 'A comparison of three-dimensional electromagnetic and RC parasitic extraction analysis of mm-wave on-chip passives in SiGe BiCMOS low-noise amplifiers', *Int. J. RF Microw. Comput.-Aided Eng.*, vol. 30, no. 2, e22019, 2020, doi: 10.1002/mmce.22019.

[4] U. Johannsen, A. B. Smolders, R. Mahmoudi, and J. A. G. Akkermans, 'Substrate loss reduction in antenna-on-chip design', in *2009 IEEE Antennas and Propagation Society International Symposium*, 2009, no. 1, pp. 1–4, doi: 10.1109/APS.2009.5172243.

[5] Woonghwan Ryu, A. L. C. Wai, Fan Wei, and Joungho Kim, 'Applications of a 3-D field solver for on-chip and package microstrip interconnection design', in *Proceedings of 3rd Electronics Packaging Technology Conference (EPTC 2000) (Cat. No.00EX456)*, Singapore, 2000, pp. 198–203, doi: 10.1109/EPTC.2000.906373.

[6] Y. Wang, J. Liu, J. L. Li, and A. Chin, 'High-gain DR circular patch on-chip antenna based on standard CMOS technology for millimeter-wave applications', in *2014 IEEE International Workshop on Electromagnetics (iWEM)*, 2014, no. 1, pp. 159–160, doi: 10.1109/iWEM.2014.6963685.

[7] Jeng-Hau Lu, C. F. Jou, Lin-Kun Wu, Sue-Chen Liao, Pei-Ling Tseng, and Tzu-Kun Ku, 'A millimeter-wave on-chip antenna using 0.15um pHEMT process', in *2015 International Workshop on Electromagnetics: Applications and Student Innovation Competition (iWEM)*, 2015, pp. 1–2, doi: 10.1109/iWEM.2015.7365034.

[8] F. Sagouo Minko and T. Stander, 'A Comparison of BEOL Modelling Approaches in Simulating mm-Wave On-Chip Antennas', in *2019 IEEE Radio and Antenna Days of the Indian Ocean (RADIO)*, 2019, pp. 1–2.

[9] A. W. Bogalecki, M. du Plessis, P. J. Venter, I. J. Nell, and M. E. Goosen, 'Integrated optical light directing structures in CMOS to improve light extraction efficiency', in *2010 International Conference on Microelectronics*, 2010, pp. 168–171, doi: 10.1109/ICM.2010.5696106.

IOP Conf. Series: Materials Science and Engineering **766** (2020) 012002 doi:10.1088/1757-899X/766/1/012002

Bias variation detection in RF amplifiers using differential OBIST

M Ballot and T Stander

Carl and Emily Fuchs Institute for Microelectronics, Dept. EEC Engineering, University of Pretoria, Pretoria, South Africa

E-mail: max.ballot@ieee.org, tinus.stander@up.ac.za

Abstract. Oscillation-based built-in self-testing with differential power detection has been shown to be a promising method for detecting degradation in radio frequency (RF) amplifiers. We demonstrate the use of the method to detect supply voltage decrease and base current variation. The efficacy of the approach is verified using simulated results of a 2.4 GHz low-noise amplifier (LNA). It is shown that these variations can be detected by using only power detection.

1. Introduction

Oscillation-based testing (OBT) [1] is a method of circuit testing that does not require test inputs. The device under test (DUT) is switched into an astable feedback loop, and the resulting oscillation frequency (f_0) and amplitude (A) measured and compared to those of a nominal circuit, with variation off nominal values Δf_0 and ΔA interpreted as indicators of circuit faults. Oscillation-based built-in self-testing (OBIST) [2] extends on this, where f_0 and A are monitored in-system or in-circuit without external test equipment. Some advanced in OBT at radio frequency (RF) [3] and mm-wave [4] frequencies have been proposed, but these still require test equipment to monitor the output. In contrast, built-in self-testing (BIST) at RF exploits in-system power detection [5, 6, 7] with built-in sources. To make RF OBT a built-in test, it has recently been demonstrated [8] that Δf_0 variation may be detected with a frequency selective power detection, with ΔA distinguished by switching between two feedback loops. This work was expanded on in [9], where the natural frequency selectivity of the OBIST feedback loop was used as a frequency discriminator, using power detectors at the input ($\mathrm{PD_{in}}$) and output ($\mathrm{PD_{out}}$) of the feedback loop. In this way, $\Delta PD = PD_{in} - PD_{out}$ is used to detect Δf_0 thus eliminating the need for digitisation of RF signals. Amplifier gain reduction was successfully detected in this way.

In this work, we propose to extend on the work in [9], by investigating the detection of a decrease in supply voltage ($\mathrm{V_{cc}}$) and variation base current ($\mathrm{I_b}$), with the circuit proposed in [9].

2. Circuit

Fig. 1 shows the differential detection OBIST circuit for a 2.4 GHz LNA proposed in [9]. The LNA is designed using an Infineon BFP843 HBT and a MACOM MASWSS0115 switches. The small-signal characteristics of switches are included in circuit simulation and the detectors are modelled as 50 Ω terminations. $\mathrm{C_{mi,o}}$ and $\mathrm{L_{mi,o}}$ are the matching elements, with $\mathrm{L_{mi,o}}$ acting as RF chokes and $C_{mi,o}$ as RF decoupling caps as well. The OBIST circuit uses a Hartley tank, C_f and $L_{f1,2}$, in the feedback loop, which creates phase shift from the output to the input of the LNA, and simultaneously acts as a frequency discriminator for ΔPD. The switches switch the system between normal operation (feedback disconnected) and OBIST (loop connected, DUT disconnected from source and load). The power detectors at the input (PD_{in}) and output (PD_{out}) of the amplifier measure RF power and convert it to a direct current (DC) voltage suitable for low rate digitization. The power detectors also load the tank, reducing the Q-factor of the tank, which results in larger variations in f_0.

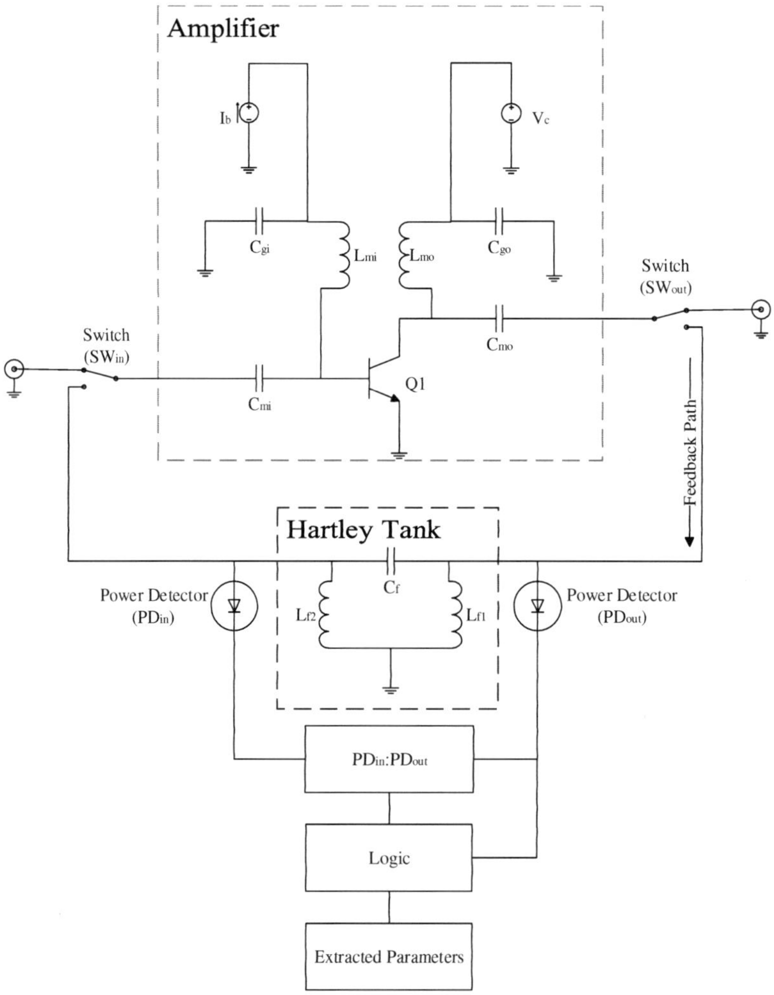

Figure 1. Differential detection OBIST schematic. [9]

3. Results

3.1. LNA performance

The LNA is simulated in Keysight ADS, using the linear S-parameter solver, to gauge the impact of OBIST circuitry on LNA performance. Fig. 2 shows that the gain decreases from 20.5 dB to 19.6 dB at 2.45 GHz, while the noise figure (NF) increases from 1.46 dB to 2 dB, as shown in Fig. 3.

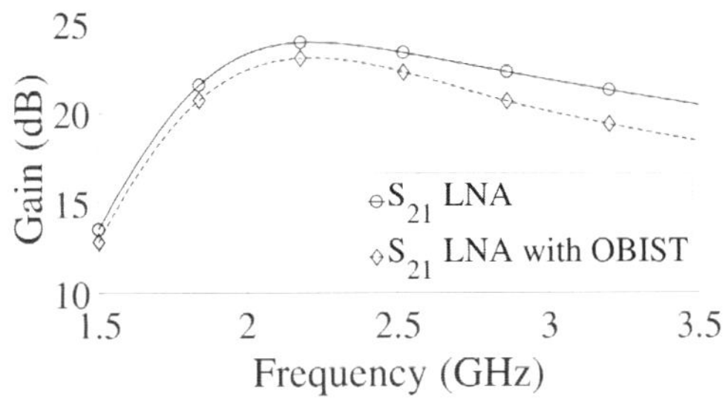

Figure 2. OBIST circuitry effect on nominal amplification.

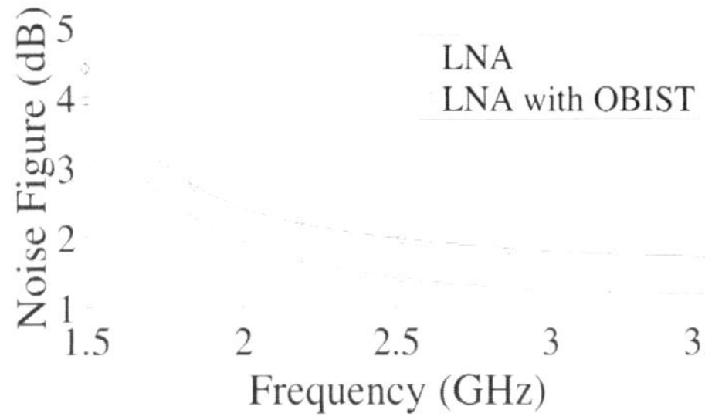

Figure 3. OBIST circuitry effect on nominal NF.

3.2. OBIST results

To extend on the investigation in gain variation on [9], Keysight ADS's harmonic balance solver (which converged in all cases with negligible residual error) was used to determine the effect of bias current I_b and bias voltage V_{cc} variation on Δf_0 and, consequently, ΔPD.

Fig. 4 shows that PD_{out} is proportional to I_b, and that there is significant variation from the nominal value at $I_b = 27$ μA. Fig. 5 shows that f_0 changes with I_b and that the ratio between the input and output power changes proportional to f_0.

Fig. 6 shows the output power compared to a decrease in V_{cc}, from the nominal value of $V_{cc} = 1.8$ V. The figure shows that an overall decrease in output power is expected, but that the decline is not monotone. Fig. 7 shows the change in f_0 due to a decrease in V_{cc}. A clear shift f_0 is evident, and is also reflected in ΔPD.

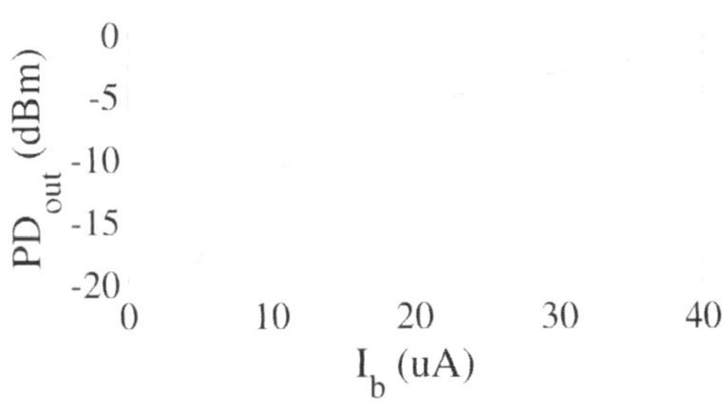

Figure 4. Variation in power at output detector due to I_b variation.

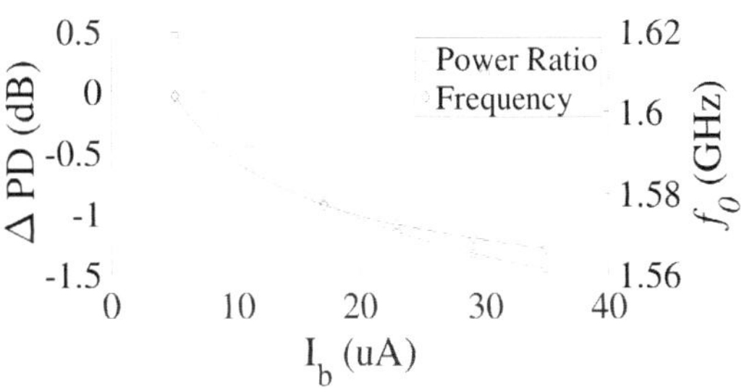

Figure 5. f_0 variation and differential detection due to I_b variation.

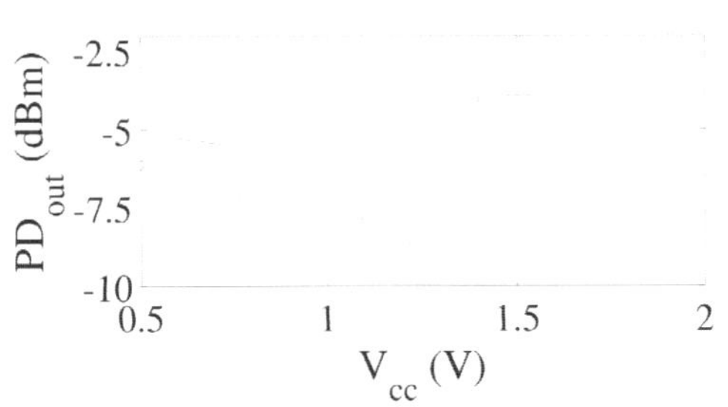

Figure 6. Decrease in power at output detector due to V_{cc} reduction.

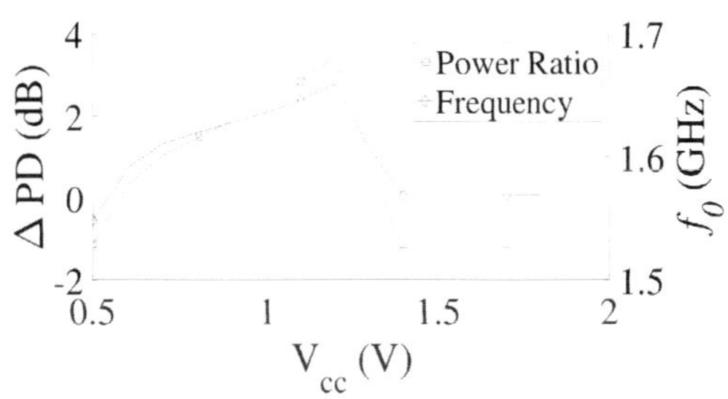

Figure 7. f_0 variation and differential detection due to V_{cc} reduction.

4. Conclusion

We extend on the gain variation detection results in [9] to show that differential power detection OBIST can be used to detect variation in V_{cc} and I_b, with minimal effect on LNA performance. The results show a direct relationship between I_b variation, the oscillator's output power, and power ratio of the power detectors. This indicates that a simple threshold detection on PD_{out} is sufficient for diagnosis. Detection of ΔI_b is critical as it significantly affects gain, noise figure, and linearity of the LNA [10]. A decrease in V_{cc} can also be detected with threshold diagnosis, though without an indication of the severity. This lack of severity detection is due to the fact that neither Δf_0, ΔPD, nor PD_{out} vary monotonically with V_{cc}, therefore more complex diagnostics (such as machine learning classifiers using multiple oscillator circuit features) have to be applied. V_{cc} on its own does not have as significant an effect on LNA performance as I_b has, because it only significantly affects compression, and only affects gain as it approaches $V_{be(on)}$ [10, 11]. This is expected to be fully detectable with simple threshold detection on PD_{out}. Future work will validate this approach experimentally, and extend this analysis to broadband multi-tone oscillation testing, as well as testing of full receiver chains. The effect of process variation on fault detection will also be investigated, as demonstrated in [8].

References

[1] Arabi K and Kaminska B 1997 *IEEE Transactions on Computer-Aided Design of Integrated Circuits and Systems* **16** 745–753 ISSN 02780070
[2] Arbet D, Stopjakova V, Majer L, Gyepes G and Nagy G 2013 *IEEE Transactions on Nanotechnology* **12** 486–497 ISSN 1536-125X
[3] Goyal A, Swaminathan M, Chatterjee A, Howard D C and Cressler J D 2012 *IEEE Transactions on Very Large Scale Integration (VLSI) Systems* **20** 1835–1848 ISSN 1063-8210
[4] Margalef-Rovira M, Barragan M J, Sharma E, Ferrari P, Pistono E and Bourdel S 2018 An oscillation-based test technique for on-chip testing of mm-wave phase shifters *2018 IEEE 36th VLSI Test Symposium (VTS)* (IEEE) pp 1–6 ISBN 978-1-5386-3774-6
[5] Lahbib I, Doukkali M A, Descamps P, Gamand P, Kelma C and Tesson O 2014 *International Journal of Microwave and Wireless Technologies* **6** 195–200 ISSN 1759-0787
[6] Xiaohua Fan, Onabajo M, Fernandez-Rodriguez F, Silva-Martinez J and Sanchez-Sinencio E 2008 *IEEE Transactions on Circuits and Systems I: Regular Papers* **55** 1794–1804 ISSN 1549-8328
[7] Ryu J Y, Kim B and Sylla I 2006 *IEEE Transactions on Instrumentation and Measurement* **55** 381–388 ISSN 0018-9456
[8] Nel H P, Stander T and Dualibe F C 2018 Built-In Oscillation-Based Self-Testing of a 2.4 GHz LNA in 0.35um CMOS *2018 25th IEEE International Conference on Electronics, Circuits and Systems (ICECS)* (Bordeaux, France: IEEE) pp 837–840 ISBN 978-1-5386-9562-3
[9] Ballot M and Stander T 2019 Built-in Oscillation-Based Testing of RF Amplifier Gain using Differential Power Detection *2019 IEEE Radio and Antenna Days of the Indian Ocean (RADIO)* (Réunion, France: IEEE) pp 1–2
[10] Liang Q, Niu G, Cressler J D, Taylor S and Harame D L 2002 Geometry and bias current optimization for SiGe HBT cascode low-noise amplifiers *IEEE Radio Frequency Integrated Circuits Symposium, RFIC, Digest of Technical Papers* pp 407–410 ISSN 0149645X
[11] Niu G, Liang Q, Cressler J D, Webster C S and Harame D L 2001 *IEEE Transactions on Microwave Theory and Techniques* **49** 1558–1565 ISSN 00189480

60 GHz Low-Noise Amplifier for Detection Systems

Pape Sanoussy Diao

Université Paris-Est, ESYCOM (UMR 9007), CNAM, CNRS, ESIEE Paris, Université Paris-Est Marne-la-Vallée, F-77454 Marne-la-Vallée, France

E-mail: pape-sanoussy.diao@esiee.fr

Abstract. This paper presents a 60 GHz low-noise amplifier (LNA) to improve detection systems performance. The considered system is based on impulse radar technology. The LNA is designed in GaAs metamorphic High Electron Mobility Transistor (m-HEMT) technology. Three stages common source transistors are used with inductive degeneration for a good trade-off between gain and noise. The post-layout simulation results show a noise factor of 2.06 dB and a gain of 14.5 dB at 60.2 GHz, for a power consumption of 13.5 mW. The simulated non-linear characteristics show an IP_{1dB} (Input 1 dB compression Point) of -10 dBm and an $IIP3$ (Input third-order Intercept Point) of -4.4 dBm. The LNA occupies an area of 1.56×1.29 mm^2.

1. Introduction

With the evolution of standards and technologies, the millimeter-wave frequencies are attracting more and more interest. The availability of unlicensed bandwidth around 60 GHz in several regions of the world can facilitate frequency harmonization and massive deployment of systems. The use of large bandwidth allows to achieve high data rates (several gigabits/second at a few meters) for communication systems and high range resolutions (a few cm) for detection systems. These opportunities create a need for broadband, low noise, low power consumption and small size amplifiers to improve systems performance.

This study is part of the development of an ultra-wideband (UWB) millimeter wave short-range detection system. Monostatic radar is considered (figure 1), with two co-located antennas and a cylindrical target of radius r and height h, located at a range R. The incidence angle θ is determined by the orientation of the target with respect to the antennas boresight.

The rest of this paper is structured in four main parts. The detection principle and LNA specifications are presented in Section 2. Section 3 details the description of the technology used and the design of the 60 GHz LNA. The layout and the electromagnetic simulation results are presented in Section 4. And finally, a conclusion and future work are presented in Section 5.

2. Principle of detection and LNA specifications

The detection principle is based on the impulse technique, using a dual-band approach [1], to improve the angular detection coverage of the target. Frequencies around 60 GHz were chosen for spectrum availability, but also for short wavelengths to detect small objects. Due to the frequency dependence of the radar cross section (RCS), we use frequency diversity [2], to maximize the received radar echo according to the target orientation angle θ.

The choice of frequencies is based on the UWB millimeter wave standardization, but also on the dimensions of the targets to be detected. It has been optimized for metal plates and cylinders whose largest dimension doesn't exceed 10 cm. In addition, it is based on the RCSs of targeted objects, which are determined using the physical optics method [3].

The architecture associated to the detection principle is shown in figure 1 for a dual-band system. Its operation consists in simultaneously transmitting pulses in both frequency bands and recombining their echoes in reception. More details are given in [1], including DSPA (Differential Structure Power Amplifier) part. This architecture can be extended to N frequency bands.

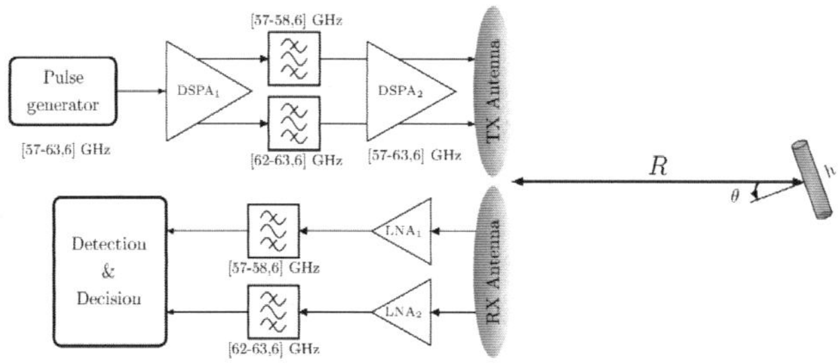

Figure 1. Example of dual band detection architecture.

The design of the system is based on the radar equation [4], which can be in the form:

$$R^4_{max} = \frac{P_t G^2 \sigma \lambda^2}{(4\pi)^3 \cdot kT\Delta f \cdot F \cdot SNR_r} \quad (1)$$

where P_t is the transmitted power, G the antenna gain, σ the RCS of the target, k the Boltzmann constant, T the temperature, Δf the receiver bandwidth, F its noise factor and SNR_r the signal-to-noise ratio required to ensure detection. By setting the objective to detect a cylinder ($r = 0.6$ cm; $h = 5.4$ cm) up to 2 m at $\theta = 0°$, we will determine the specifications of the LNAs. To do this, we consider a quad-band system in 57-66 GHz, around the frequencies 57.8 GHz, 60.2 GHz, 62.8 GHz and 65.2 GHz. The output power of each channel of the DSPA$_2$ is set at 15 dBm, taking into account the standardization and antennas gain (12 dBi). The bandwidth of each band is 1.6 GHz and the filter losses are set to 3.5 dB [5]. For a single pulse, the SNR_r for a detection probability of 90% and a false alarm probability of 10^{-6} is 13.2 dB for a non-fluctuating target. With a non-coherent integration of 4 pulses in the 4 frequency bands, the SNR_r is 8.3 dB for the same probabilities. Based on non-coherent integration and using equation (1), the specifications of the LNAs are at least 12,5 dB of gain and at most 3 dB of noise factor.

3. Technology description and circuit design

The 60.2 GHz LNA is designed in OMMIC's D007IH technology, which has a 70 nm gate length. This technology has a f_T/f_{max} of 300 GHz/450 GHz. It offers a depletion transistor m-HEMT with $V_{DS_{max}}$ of 3 V and $I_{DSS_{max}}$ of 400 mA/μm. The choice of this technology is based in part on its low losses, 0.22 dB/mm at 60 GHz. The technology process consists of a 3.5 μm metal on its underside, a 100 μm thick GaAs substrate above which there are different metal layers.

The LNA is designed with Keysight Advanced Design System (ADS). It consists of a three stages structure using identical transistors. The size of the transistor (2 × 25 μm of gate) is chosen for a better trade-off between gain and noise [6]. The optimal bias point corresponds to a voltage $V_{DS} = 1$ V for an I_{DS} current of about 4.1 mA. Unconditional stability has been ensured both by inductive degeneration of the source, but also by the use of resistors in the bias circuits. The first two stages are almost identical and are matched with a good trade-off between gain and noise. Matching of the third stage is optimized in gain because its influence is less significant on the whole structure noise factor.

Figure 2 presents the schematic of the LNA. All bias circuits are made with quarter-wave transmission lines. They include GaAs implanted resistors (R_D and R_G) and bypass capacitors C_D. The degeneration of the transistor source is made by the transmission line TL$_S$. C_1, TL$_1$

IOP Conf. Series: Materials Science and Engineering **766** (2020) 012003 doi:10.1088/1757-899X/766/1/012003

and TL_2 form the input matching network. C_2, TL_3 and C_3 on the one hand, and C_4, TL_4 and C_5 on the other hand, form the inter-stages matching networks between stages 1 and 2 and stages 2 and 3 respectively. The output maching network is formed by C_6 and TL_5.

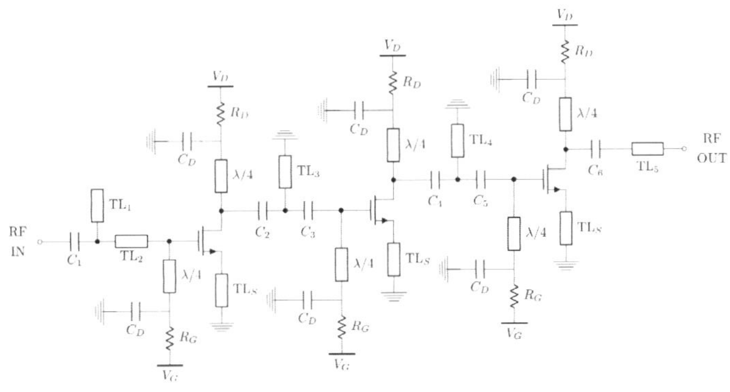

Figure 2. Schematic of the LNA design.

4. Layout and simulation results

The layout of the LNA is shown in figure 3 and occupies 1.56×1.29 mm^2. It includes RF (Radio Frequency) and DC (Direct Current) pads. All ground connections are made by vias holes.

The post-layout simulation results with ADS Momentum Microwave solver are presented in figures 4-6. The circuit is unconditionally stable up to 200 GHz as shown in figure 4.

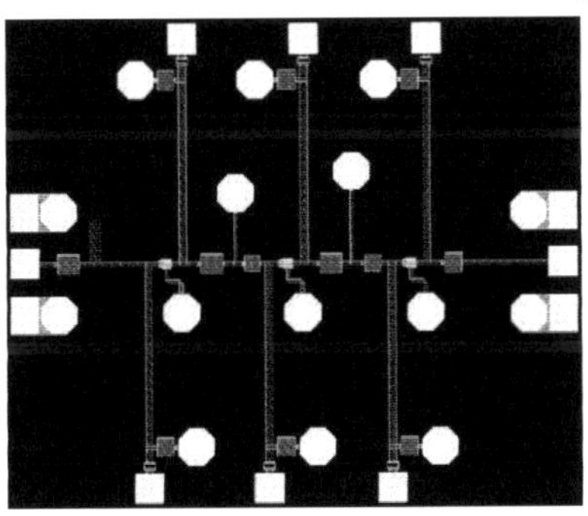

Figure 3. Layout of the LNA.

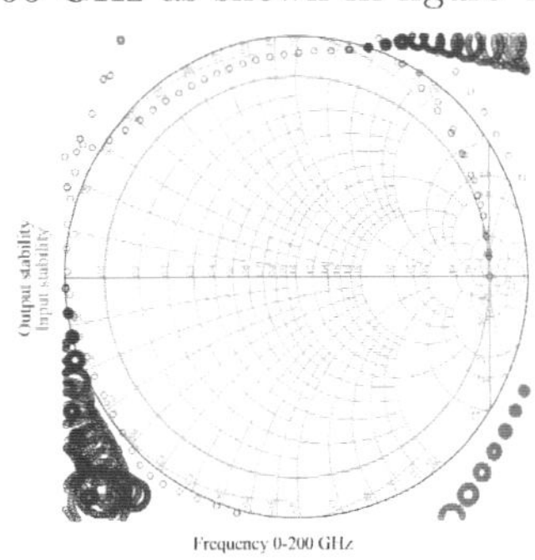

Figure 4. Stability of the LNA.

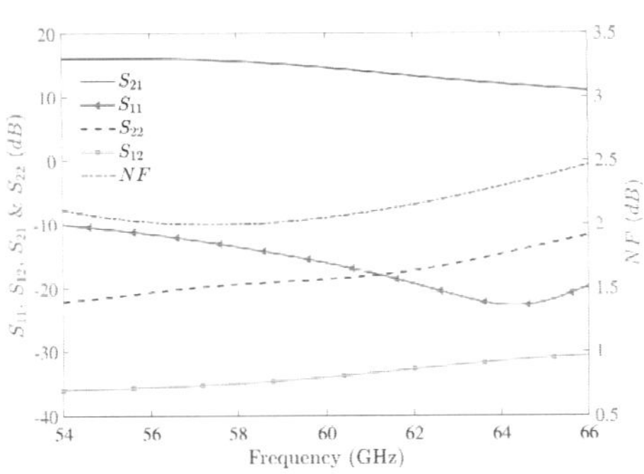

Figure 5. S-parameters of the LNA.

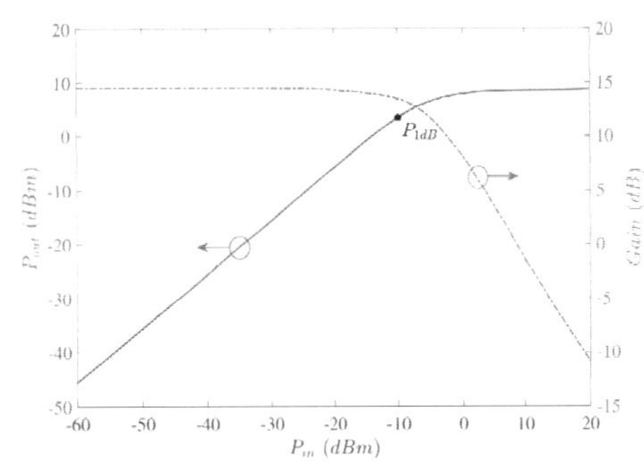

Figure 6. Gain and P_{1dB}.

Our LNA achieves a gain of 14.5 dB and a noise factor 2.06 dB at 60.2 GHz. The reflection coefficients S_{11} and S_{22} are less than -15 dB in the band of interest 59.4-61 GHz. In this band,

IOP Conf. Series: Materials Science and Engineering **766** (2020) 012003 doi:10.1088/1757-899X/766/1/012003

we observed a gain ripple of 0.9 dB and a noise factor ripple less than 0.1 dB. The non-linear characteristics of the LNA present an IP_{1dB} of -10 dBm (figure 6) and an $IIP3$ of -4.4 dBm. The performance of the designed LNA compared to other LNAs in the state-of-the-art are given in table 1. Compared to [7] and [8], our LNA presents a better noise factor, a good gain with a moderate power consumption and occupies less space. Its non-linear characteristics are much better than those of references [8]- [10]. Note that CMOS technologies are generally more suitable for miniature and low-power consumption components compared to HEMT technologies.

Table 1. LNA position in the state-of-the-art.

Ref.	Technology	Freq. (GHz)	Gain (dB)	NF (dB)	P_{DC} (mW)	Area (mm^2)
[7]$^+$	50 nm GaAs*	50-90	27	2.6	45	2.3×1.6 mm^2
[8]$^+$	100 nm GaAs*	60-90	19	2.5	56	3.5×1.0 mm^2
[9]$^+$	40 nm CMOS	60	12.5	3.8	20.4	0.63×0.31 mm^2
[10]$^+$	65 nm CMOS	60	20.2	5.2	28	0.54×0.80 mm^2
This Work	70 nm GaAs*	60.2	14.5	2.06	13.5	1.56×1.29 mm^2

$^+$ Measures, * m-HEMT

5. Conclusion and future work

A 60 GHz LNA designed in GaAs m-HEMT technology was presented. This design was done for a multi-band system using a frequency diversity to improve detection coverage. Our designed LNA offers good performance with only 2.06 dB noise factor at 60.2 GHz. With three stages, it achieves a gain of 14.5 dB, while consuming 13.5 mW. With this performance, the range and overall detection coverage are improved compared to conventional single-band system.
As future work, we can mention the realization and characterization of the designed LNA.

References

[1] P. S. Diao, T. Alves, M. Villegas, and B. Poussot, Compact millimeter wave architecture dedicated to object detection using dual band-dual polarization and impulse method, in Proceedings of the 13th *Conference on Ph.D. Research in Microelectronics and Electronics* (*PRIME*), Jun. 12-15, 2017, Giardini Naxos, Italy. IEEE, Jul. 2017, pp. 161-164.

[2] D. K. Barton, *Frequency Diversity Theory* Artech House Inc, Ed., 1977, Radars vol 6, Section 2, pp. 35-114, in *Frequency Agility and Diversity.*

[3] E. F. Knott, J. F. Shaeffer, and M. T. Tuley, *High-Frequency RCS Prediction Techniques* 2nd ed., SciTech Publishing, Inc, Ed., Raleigh, NC (USA), 2004, Chapter 5, pp 183–224, in *Radar Cross Section.*

[4] M. I. Skolnik, *The Radar Equation* 2nd ed., McGraw Hill, Inc, Ed., 1980, Chapter 2, pp. 15-67, in *Introduction to Radar Systems.*

[5] R. Abdaoui, M. Villegas, G. Baudoin, and A. Diet, Microstrip band pass filter bank for 60 GHz UWB impulse radio multi band architectures, in Proceedings of the *IEEE MTT-S International Microwave Workshop Series on Millimeter Wave Integration Technologies*, Sep. 15-16, 2011, Sitges, Spain. IEEE, Oct. 2011, pp. 192-195.

[6] P. S. Diao, T. Alves, B. Poussot, and M. Villegas, A 60 GHz Low-Noise Amplifier for Detection Systems, in Proceedings of the *2019 IEEE Radio and Antenna Days of the Indian Ocean* (*RADIO*), Sep. 23-26, 2019, Saint-Gilles-Les-Bains, Reunion Island. IEEE, Jan. 2020, pp. 1-2.

[7] P. M. Smith *et al.*, A 50 nm MHEMT millimeter-wave MMIC LNA with wideband noise and gain performance, in Proceedings of the *IEEE MTT-S International Microwave Symposium* (*IMS2014*), Jun. 1-6, 2014, Tampa, FL, USA. IEEE, Jul. 2014, pp. 1-4.

[8] A. Bessemoulin, J. Grunenputt, P. Felton, A. Tessmann, and E. Kohn, Coplanar W-band low noise amplifier MMIC using 100-nm gate-length GaAs PHEMTs, in Proceedings of the 34th *European Microwave Conference*, Oct. 12-14, 2004, Amsterdam, The Netherlands, vol. 1. IEEE, 2005, pp. 25-28.

[9] H. Gao et *al.*, A 48–61 GHz LNA in 40-nm CMOS with 3.6 dB minimum NF employing a metal slotting method, in Proceedings of the *IEEE Radio Frequency Integrated Circuits Symposium* (*RFIC*), May 22-24, 2016, San Francisco, CA, USA. IEEE, Jul. 2016, pp. 154-157.

[10] C. So and S. Hong, 60 GHz variable gain LNA with small NF variation, in Proceedings of the *IEEE International Symposium on Radio-Frequency Integration Technology* (*RFIT*), 30 Aug.1 Sep., 2017, Seoul, South Korea. IEEE, Sep. 2017, pp. 171-173.

Deep Fading in a Reverberation Chamber for Wireless Device Testing

Christoph Cammin, Dmytro Krush, Ralf Heynicke and Gerd Scholl
Electrical Measurement Engineering, Helmut-Schmidt-University Hamburg, Germany
E-mail: `christoph.cammin@hsu-hh.de`

Abstract. Reverberation chambers (RCs) are a feasible option as test environment for evaluation of wireless devices with integrated antenna as they inherently emulate a Rayleigh fading channel. In this paper the dependency of deep fading on the frequency and position/orientation of an equipment under test (EUT) antenna is evaluated for different numbers of RC steps and loading conditions. The measured data were applied to an exemplary standard for wireless communication. The measurement results are highly repeatable and independent of frequency or position/orientation of the EUT and loading conditions.

1. Introduction

Reverberation chambers (RCs) are used in various standards as reproducible test environments, e.g. for electromagnetic compatibility testing [1], or mobile phone performance testing [2, 3, 4]. They consist of an electromagnetically shielded volume in which mode stirrers are built in. The latter may be mechanically moving or rotating metal plates or differently designed facilities in order to change the electromagnetic field distribution within the volume. A sequence consisting of a certain number of individual measurements is taken at various mode stirrer conditions (steps). Over a full measurement sequence an equipment under test (EUT) is exposed to a statistically quasi-homogeneous and quasi-isotropic electromagnetic field with a Rayleigh-distributed field strength, i.e., Rayleigh fading [2, 3, 5, 6]. Due to these characteristics a RC is an auspicious test environment for wireless devices with integrated antennas, especially if the mounting orientation and therefore the direction and polarization of the incident electromagnetic field is unknown, such as wireless sensors in industrial environments.

IO-Link Wireless (IOLW) is a new standard for wireless factory automation [7, 8]. The physical layer is based on Bluetooth Low Energy within the 2.4 GHz band [7]. RCs are suggested as test environment for IOLW [9] and thus IOLW is regarded as reference system as already elaborated in [10].

2. Measurement Setup

The measurements were performed in a Bluetest RTS60 RC. The inner dimensions are approx. 1.8 m (height), 1.2 m (depth) and 1.9 m (length). That type of RC utilizes two planar mode stirrers, a turntable for the EUT and three switchable, fixed and orthogonally arranged antennas for excitation (see e.g., [3]). The measurement sequences consist of 300, 600 and 1500 steps (i.e., mode stirrer and fixed antenna combinations). A sleeve reference dipole antenna was used as EUT. The S-parameters between the EUT antenna and the fixed antenna arrangement were measured for 1001 frequencies within 2.4 to 2.5 GHz. In order to adapt the power delay profile (PDP) of the RC to a large variety of industrial environments, commercially available pyramidal and cuboid foam absorbers were placed in the RC (see e.g., [6, 11, 12]). The absorbers were at least one wavelength away from the next point of the EUT. The resulting RMS (root mean square)-delay spread τ_{RMS} of the power delay profile was about 190 ns, 50 ns and 23 ns, respectively. The EUT antenna was mounted on three radii on the turntable and each three different heights and each three orthogonal orientations (i.e., polarizations: radial, tangential, vertical), as shown in Figure 1. Thus in total 27 measurement sequences were recorded for each number of steps and loading or τ_{RMS}, respectively.

The measurements confirm that the mean S_{21}-parameters over the stirrer combinations for the 27 measurement sequences are within the same order of magnitude and almost flat over the frequency range for each loading case (an example is shown in [10, Figure 1]). Therefore, these mean values can be assumed to be equal in good approximation and in this case it is irrelevant what is averaged over. Figure 2 shows exemplarily the empirical cumulative distribution (eCDF) of fading for the 27 measurement sequences at 2.44 GHz with 300 Steps and $\tau_{RMS} \approx 50$ ns.

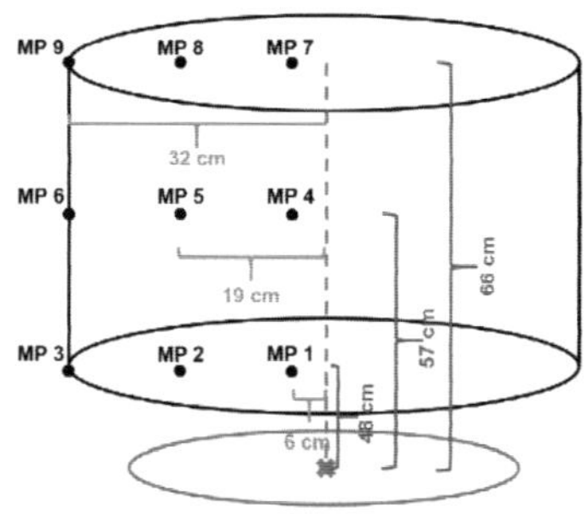

Figure 1. Measurement points (MPs) on the turntable.

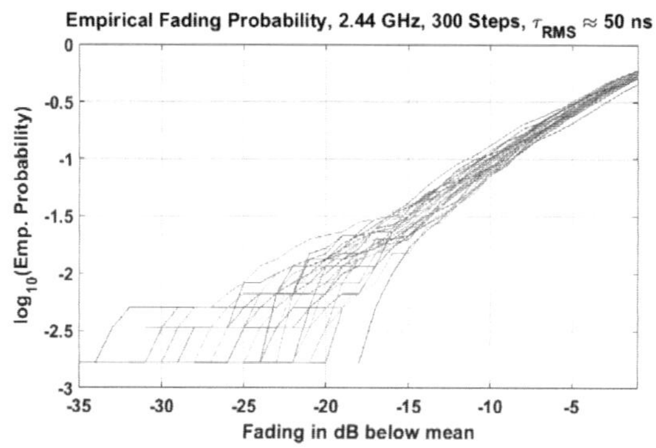

Figure 2. Fading eCDF for the 27 positions/orientations of the EUT, exemplarily shown for 300 steps and $\tau_{RMS} \approx 50$ ns.

3. Evaluation for a limited Subset of Frequencies

For wireless device evaluation, the probability of fading (below the mean value of the radio channel) is of importance, as these can lead to outages. Thus a constant probability, independent of the frequency and position/orientation of the EUT is a key requirement. In the following sections, the RC is evaluated on a subset of 73 frequencies from 2403 MHz to 2475 MHz with 1 MHz spacing. This corresponds to the IOLW standard.

3.1. Rayleigh Fading Validation

In [4, Appendix C] a procedure for Rayleigh fading evaluation based on a χ^2-test is described. This procedure was performed for each loading, each EUT position/orientation, each stirring sequence and each frequency used by IOLW. (According to [4], here 11 vector network analyzer (VNA) frequency steps were used for the frequency averaging, which corresponds to presumed 1.1 MHz signal bandwidth.) The Rayleigh fading hypothesis was never rejected for the unloaded RC. The rejection probability increases for a loaded RC, but remained below 1%, even in the worst case of the heaviest loaded RC. Nevertheless, the described procedure in [4] uses 15 bins, whereby the bin which includes the samples with the lowest received power covers $-\infty$ dB to -11.6 dB (relative to the mean). Therefore all samples which represents deep fading are covered only by that single bin. In the following, the deep fading range is examined in more detail, since a fading of -10 dB can often be handled by the system without errors, but typically much deeper fading cannot.

3.2. Dependence of Fading on the Frequency

The empirical probability of measured S_{21}-values whose magnitudes are at least -10 dB, -20 dB, and -30 dB relative to the mean S_{21}-value as a function of frequency are calculated. Figure 3 show the mean, minimum, maximum and the (normalized) estimated standard deviation (SD) according to [1] over the $N = 27$ positions/orientations of the EUT antenna of the S_{21}-parameters exemplarily for 300 and 600 steps and $\tau_{RMS} \approx 50$ ns.

The mean over the 27 positions/orientations is almost frequency flat and in accordance with theory of Rayleigh fading. The SD increased with deeper fading. Additionally, for some frequencies there is at least one position/orientation, at which no fading of more than 20 dB below the mean value is observed. Table 1 lists the SDs (as shown in Fig. 3) averaged over the 73 frequencies and normalized to the mean value over the frequencies and positions/orientations for -10 dB, -20 dB, and -30 dB and each, 300, 600 and 1500 steps, respectively. This normalized SD values decreases with a higher number of steps, but is almost independent of τ_{RMS}.

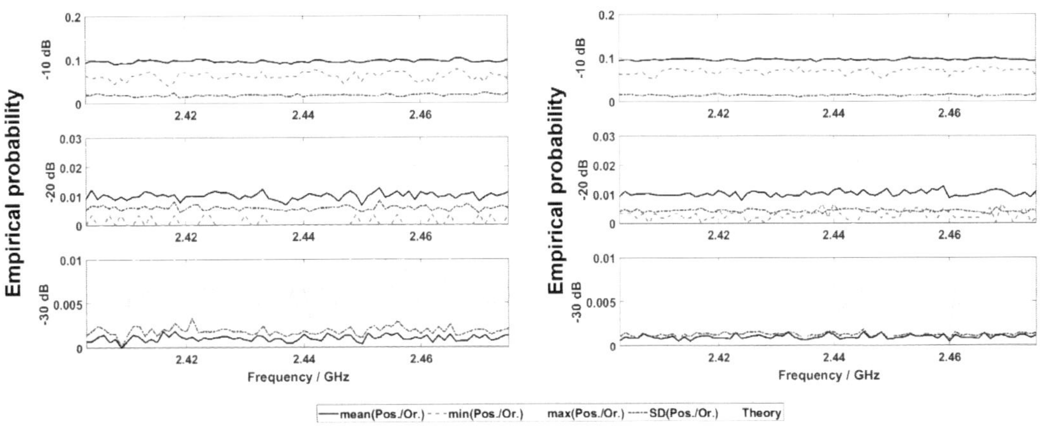

Figure 3. Empirical probability of fading as a function of frequency for 300 (left) and 600 (right) steps (both for $\tau_{RMS} \approx 50$ ns).

3.3. Dependence of Fading on the Position or Orientation

The dependency of the fading as a function of the position/orientation of the EUT antenna is evaluated in this section. Figure 4 shows the mean, minimum, maximum and the SD of the 73 frequencies for each position/orientation exemplarily for $\tau_{RMS} \approx 50$ ns and both, 300 and 600 steps. The results are similar to the previous ones. The SDs (as shown in Fig. 4) averaged over the positions/orientations and normalized to the total mean value over the frequencies and positions/orientations are listed in Table 1. The normalized SD values averaged over the positions/orientations are the same as the ones which are averaged over the frequencies. The mean values over all frequencies and positions/orientations for the specific number of steps and loading fit very well with the theory of Rayleigh fading.

4. Summary

The measurements show that the averaged empirical probability of fading is independent of frequency or the position/orientation of the EUT and also independent of loading or τ_{RMS}, respectively. However, for some individual combinations of position and frequency, very deep fading did not occur. As the SD generally decreases with increasing number of (independent) steps, the number of this outliers can be reduced. Overall, the RC is a reproducible test environment for wireless device performance under Rayleigh fading conditions.

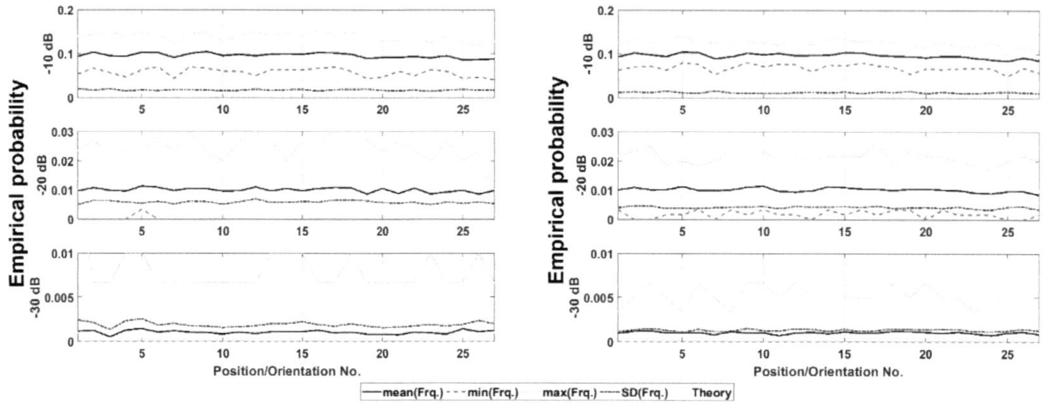

Figure 4. Empirical probability of fading as a function of position and orientation of the EUT for 300 (left) and 600 (right) steps (both for $\tau_{RMS} \approx 50$ ns).

Table 1. Mean values and SD for different fading levels, steps and τ_{RMS}. The theoretical values from Rayleigh fading for the mean are given in italics.

Fading depth	τ_{RMS}	300 Steps 190 ns	300 Steps 50 ns	300 Steps 23 ns	600 Steps 190 ns	600 Steps 50 ns	600 Steps 23 ns	1500 Steps 190 ns	1500 Steps 50 ns	1500 Steps 23 ns
-10 dB	Mean (*0.0952*)	0.0948	0.0958	0.0968	0.0947	0.0959	0.0978	0.0949	0.0956	0.0973
-10 dB	SD Freq.	0.18	0.19	0.19	0.13	0.14	0.14	0.08	0.10	0.11
-10 dB	SD Pos./Or.	0.18	0.18	0.18	0.13	0.13	0.13	0.08	0.09	0.09
-20 dB	Mean (*0.0100*)	0.0100	0.0103	0.0100	0.0098	0.0101	0.0103	0.0099	0.0100	0.0101
-20 dB	SD Freq.	0.56	0.58	0.56	0.41	0.41	0.40	0.25	0.26	0.27
-20 dB	SD Pos./Or.	0.57	0.58	0.56	0.41	0.41	0.40	0.25	0.26	0.26
-30 dB	Mean (*0.0010*)	0.0010	0.0010	0.0010	0.0010	0.0010	0.0011	0.0010	0.0010	0.0010
-30 dB	SD Freq.	1.80	1.79	1.77	1.25	1.28	1.23	0.80	0.82	0.80
-30 dB	SD Pos./Or.	1.83	1.83	1.79	1.25	1.29	1.24	0.81	0.82	0.80

References

[1] IEC 61000-4-21:2011
[2] Holloway C L, Hill D A, Ladbury J M, Wilson P F, Koepke G and Coder J 2006 *IEEE Trans. on Antennas and Propagation* **54** 3167–3177 ISSN 0018-926X
[3] Kildal P S, Chen X, Orlenius C, Franzen M and Patane C S L 2012 *IEEE Trans. on Antennas and Propagation* **60** 3875–3891 ISSN 0018-926X
[4] 3GPP TR 37.977 V15.0.0 (2018-09)
[5] Kostas J G and Boverie B 1991 *IEEE Trans. on Electromagnetic Compatibility* **33** 366–370 ISSN 0018-9375
[6] Genender E, Holloway C L, Remley K A, Ladbury J, Koepke G and Garbe H 2008 *2008 Int. Symp. on Electromagnetic Compatibility - EMC Europe* pp 1–6
[7] IO-Link Community 2018 IO-Link Wireless System Extensions - Specification Version 1.1, March 2018, Order No: 10.112
[8] Heynicke R, Krush D, Cammin C, Scholl G, Kaercher B, Ritter J, Gaggero P and Rentschler M 2018 *Journal of Sensors and Sensor Systems* **7** 131–142
[9] Cammin C, Krush D, Heynicke R and Scholl G 2018 *Journal of Sensors and Sensor Systems* **7** 183–192
[10] Cammin C, Krush D, Heynicke R and Scholl G *IEEE RADIO 2019* 23rd to 26th Sept. 2019, Reunion Island
[11] Remley K A, Dortmans J, Weldon C, Horansky R D, Meurs T B, Wang C M, Williams D F, Holloway C L and Wilson P F 2016 *IEEE Transactions on Electromagnetic Compatibility* **58** 661–672 ISSN 0018-9375
[12] Cammin C, Krush D, Heynicke R and Scholl G 2018 *tm - Technisches Messen* **85** 420–433

AUTHOR INDEX

Institute of Physics
Dirac House, Temple Back
Bristol BS1 6BE UK

ISSN: 1757-8981
ISBN 978-1-7138-2664-4

CURRAN ASSOCIATES INC.
proceedings.com